Human Anatomy

About this Book

Instructors and students reviewing, adopting, or adapting this textbook are encouraged to provide feedback (or report errors) online at: https://bit.ly/feedback-human-anatomy.

Human Anatomy
Self-Assessment Review Questions

Michael F. Nolan
John P. McNamara

Virginia Tech Carilion School of Medicine
in association with Virginia Tech Publishing
Blacksburg, VA

A free version of this book can be downloaded from:
https://doi.org/10.21061/human-anatomy-self-assessment

First published in 2022 by the Virginia Tech Carilion School of Medicine in association with Virginia Tech Publishing

Virginia Tech Carilion School of Medicine
2 Riverside Circle
Roanoke, VA 24016 USA

Virginia Tech Publishing
University Libraries at Virginia Tech
560 Drillfield Drive
Blacksburg, VA 24061 USA

Cataloging-in-Publication Data
Names: Nolan, Michael F., author | McNamara, John P., author.
Title: Human Anatomy: Self-Assessment Review Questions / Michael F. Nolan and John P. McNamara.
Description: Roanoke, VA : Virginia Tech Carilion School of Medicine in association with Virginia Tech Publishing, 2022. | Summary: "The anatomy questions included here are formatted in the short answer or fill in the blank style that requires the retrieval rather than the recognition of information; a format more closely aligned with that needed for everyday clinical practice. Each chapter includes a series of questions related to a particular area of anatomy. Answers to those questions are included at the end of the book." – Preface.
Identifiers: ISBN: 978-1-957213-30-9 (PDF) | ISBN: 978-1-957213-97-2 (pbk) | DOI: https://doi.org/10.21061/human-anatomy-self-assessment
Subjects: HUMAN ANATOMY / Laboratory manuals. | HUMAN ANATOMY / Problems and Exercises.

Table of Contents

Preface

Human anatomy is one of the foundational disciplines in the training of health care professionals. Knowledge of human structure at both the macroscopic and microscopic levels serves as a framework for the study of normal function and dysfunction.

Resources used by students in their efforts to learn about human anatomy are exceptionally diverse in terms of scope, depth and focus as well as in the modalities and pedagogical approaches used. Diversity also exists in how knowledge acquisition is measured and evaluated.

Many students are familiar with formative and summative tests structured in the multiple-choice format. Most high stakes exams utilize this format and not surprisingly, numerous self-assessment resources are available to help students prepare for these types of examinations.

The self-assessment questions presented here are intended to help students in a different setting, a setting such as a ward, the clinic or the operating room where questions from teachers and preceptors are not likely to be presented as recognition exercises, that is, in the multiple-choice format, but rather as recall tasks. When information is obtained from patients in a teaching setting, it is not typical for a teacher to pose a question by asking "Which of the following five conditions does the patient have?" More likely the student will be asked to supply answer or explanation to an open-ended inquiry, such as, "What arteries do you think we will have to ligate in order to remove this inflamed appendix safely?" In this situation, the student will have to recall information or facts from previous experience or learning. Educators commonly describe this behavior as demonstrating knowledge by recall rather than recognition.

The anatomy questions included here are formatted in the short answer or fill in the blank style that requires the retrieval rather than the recognition of information; a format more closely aligned with that needed for everyday clinical practice.

Each chapter includes a series of questions related to a particular area of anatomy. Answers to those questions are included at the end of the book. It is our belief that this alternate approach to self-assessment in basic human anatomy will be helpful to students in preparing for clinical-based learning activities as well as for other types of knowledge assessment and evaluation.

MFN
JPM

Part 1: Self-Assessment Questions

1 Back and Spine

1. Indicate the number of vertebrae that comprise each region of the vertebral column.

   a. cervical ______________________

   b. thoracic ______________________

   c. lumbar ______________________

   d. sacral ______________________

2. Which two (2) regions of the vertebral column are characterized by curvatures in the sagittal plane with a concavity directed posteriorly?

   a. ______________________

   b. ______________________

3. Name each of the ligaments of the vertebral column described below.

   a. courses vertically along the anterior surfaces of adjacent vertebral bodies

   ____

   b. courses between adjacent laminae

   ____

   c. courses between adjacent spinous processes

   ____

   d. courses vertically along the posterior surfaces of adjacent vertebral bodies

   ____

4. What is the name of the thickened supraspinous ligament of the cervical spine?

   ____

5. What is the name of the synovial joint formed by the superior and inferior articular processes of adjacent vertebrae?

____

6. Which group of vertebrae is characterized by transverse foraminae?

____

7. The carotid tubercle is an osteological feature of which vertebra?

____

8. The dens is an osteological feature of which vertebra?

____

9. Which vertebra is commonly referred to a "vertebra prominens"?

____

10. Name the two (2) parts of the vertebral arch.

a. ____

b. ____

11. What are the names of the two (2) parts of the intervertebral disc?

a. central fluid part _______________________________

b. peripheral fibrous part _______________________________

12. What is the name of the ligamentous structure that extends from the posterior arch of C1 (atlas) to the base of the cranium?

____

13. What is the name of the venous plexus located within the vertebral canal?

____

14. What is the name of the thickened connective tissue origin of the latissimus dorsi?

____

15. Name the three (3) muscles that form the borders of the triangle of auscultation.

a. _______________________________________

b. _______________________________________

c. _______________________________________

16. What two (2) muscles form the superficial layer of intrinsic back muscles?

a. _______________________________________

b. _______________________________________

17. What muscle group forms the intermediate layer of intrinsic back muscles?

____

18. What muscle group forms the deep layer of intrinsic back muscles?

____

19. What three (3) muscles form the suboccipital triangle?

a. _______________________________________

b. _______________________________________

c. _______________________________________

20. What two (2) structures lie in the floor of the suboccipital triangle?

a. artery ______________________

b. nerve ______________________

21. Which spinal nerve gives rise to the greater occipital nerve?

____

22. What vessel passes through the posterior atlanto-occipital membrane?

____

23. In the adult, what vertebral level marks the caudal extent of the spinal cord?

____

24. What is the name of the collection of dorsal and ventral roots that lie caudal to the tip of the spinal cord?

____

25. What is the name of the pia matter structure that lies caudal to the tip of the spinal cord?

____

26. What is the name of the CSF filled space located below the caudal tip of the spinal cord?

____

27. What is the name of the meningeal structure that attaches to the dorsum of the coccyx?

____

28. What is the name of the pia-derived structure that anchors the spinal cord to the lateral wall of the vertebral canal?

____

29. What are the three (3) borders of the lumbar (Petit's) triangle?

a. ____________________________________

b. ____________________________________

c. ____________________________________

30. What part of the intervertebral disc is a remnant of the embryological notochord?

____

31. What are the three (3) muscles that comprise the erector spinae (sacrospinalis) muscle group?

a. ____________________________________

b. ____________________________________

c. ____________________________________

32. What are the three (3) muscles that comprise the transversospinal group of the posterior spinal muscles?

a. ____________________________________

b. ____________________________________

c. ____________________________________

33. What are the four (4) components of the occipitoaxial ligament?

a. ______________________________

b. ______________________________

c. ______________________________

d. ______________________________

34. What nerve innervates the muscles of the suboccipital triangle?

____

35. What is the name of the lower sacral segments of the spinal cord?

____

36. What vertebral segment marks the caudal extent of the spinal subarachnoid space?

____

2 Upper Limb

1. Name the two (2) parts of the clavipectoral fascia.

   a. ______________________________

   b. ______________________________

2. What vein lies in the deltopectoral groove?

   ____

3. What vein passes through the costocoracoid membrane?

   ____

4. What spinal segment (spinal nerve) provides sensory innervation to each of the following digits?

   a. palmar surface of the little finger ______________

   b. palmar surface of the thumb ______________

   c. palmar surface of the long finger ______________

5. What veins lie in the subcutaneous tissue of the medial surface of the forearm?

   ____

6. What vein serves as an anastomosis between the cephalic vein and the basilic vein?

   ____

7. What nerve continues into the forearm as the lateral antebrachial cutaneous nerve?

____

8. What anatomical landmark marks the point where the axillary artery becomes the brachial artery?

____

9. What landmarks are used to divide the axillary artery into three (3) parts?

____

10. Name the major branches of the axillary artery:

    a. first part ______________________

    b. second part ______________________

    ____

    c. third part ______________________

    ____

    ____

11. What is the insertion of the pectoralis minor?

____

12. What is the insertion of the pectoralis major?

____

13. What is the insertion of the serratus anterior?

____

14. What is the insertion of the trapezius?

____

15. What is the insertion of the latissimus dorsi?

____

16. What is the insertion of the subscapularis?

____

17. What is the insertion of the deltoid?

____

18. Which of the "rotator cuff" muscles is NOT an external rotator of the glenohumeral joint?

____

19. What muscle forms the medial wall of the axilla?

____

20. What anatomical feature marks the lateral wall of the axilla?

____

21. What anatomical landmark marks the point where the subclavian artery becomes the axillary artery?

____

22. What four (4) muscles form the borders of the quadrangular space?

a. ______________________________

b. ______________________________

c. ______________________________

d. ______________________________

23. What two (2) structures pass through the quadrangular space?

a. ______________________________

b. ______________________________

24. What nerve lies in the intermuscular fascia separating the biceps brachii and the brachialis?

____

25. What nerve innervates the muscles of the posterior compartment of the arm?

____

26. What nerve crosses the elbow joint on the deep surface of the brachioradialis?

____

27. What are the two (2) terminal branches of the brachial artery?

a. ______________________________

b. ______________________________

28. What is the bony insertion of the biceps brachii?

____

29. What is the insertion of the brachialis?

____

30. What vein lies superficial to the bicipital aponeurosis?

____

31. What muscles form the medial and lateral borders of the cubital fossa?

a. medial border ______________________________

b. lateral border ______________________________

32. What two (2) muscles form the floor of the cubital fossa?

a. ______________________________________

b. ______________________________________

33. What nerve enters the forearm, passing posterior to the medial epicondyle?

____

34. What are the four (4) muscles that form the superficial layer of the anterior compartment of the forearm?

a. ______________________________________

b. ______________________________________

c. ______________________________________

d. ______________________________________

35. What muscle forms the intermediate layer of the anterior compartment of the forearm?

____

36. What three (3) muscles form the deep layer of the anterior compartment of the forearm?

a. ______________________________

b. ______________________________

c. ______________________________

37. Which one (1) of the muscles of the anterior compartment is NOT innervated by the median nerve?

____

38. Which of the muscles of the anterior compartment of the forearm has a tendon that passes superficial to the flexor retinaculum?

____

39. Which of the muscles of the anterior compartment of the forearm receives motor innervation by way of two different peripheral nerves?

____

40. Which two (2) muscles of the posterior compartment of the forearm do NOT have tendons that cross the wrist joint?

a. ______________________________

b. ______________________________

41. What nerve innervates muscles that extend the wrist?

____

42. What three (3) muscles form the hypothenar eminence?

a. ______________________________

b. ______________________________

c. ______________________________

43. What four (4) muscles form the thenar eminence?

a. ______________________________

b. ______________________________

c. ______________________________

d. ______________________________

44. Which of the thenar muscles is innervated by both the median nerve and the ulnar nerves?

____

45. Which of the intrinsic muscles of the hand adduct the medial four fingers?

____

46. Which of the intrinsic muscles of the hand abduct the medial four fingers?

____

47. What nerve enters the hand by passing through Guyon's canal?

____

48. What branch of the ulnar nerve provides sensory innervation to the skin over the palmar surface of the little finger?

____

49. What is the name of the fibrocartilaginous structure that "deepens" the glenohumeral joint?

____

50. What are the two (2) parts of the coracoclavicular ligament?

a. ____

b. ____

51. Which ligament of the shoulder does NOT cross the glenohumeral joint?

____

52. Name the bones that form the proximal row of carpal bones.

a. ____

b. ____

c. ____

d. ____

53. Name the bones that form the distal row of carpal bones.

a. ____

b. ____

c. ____

d. ____

54. What two (2) structures pass through the triangular interval?

a. ____

b. ____

55. What artery passes through the triangular space?

____

56. What muscle forms the medial border of the ‘anatomical snuff box”?

____

57. What arterial pulse can be palpated in the “anatomical snuff box”?

____

58. Which of the carpal bones can be palpated in the floor of the “anatomical snuff box”?

____

59. Which joint of the hand lies immediately deep to the extensor hood (extensor expansion)?

____

60. What is the action of the lumbrical muscle on each of the following joints of the finger?

a. MP ______________________

b. PIP ______________________

c. DIP ______________________

61. Which two heads of the triceps brachii originate from the humerus?

a. ______________________

b. ______________________

62. What artery gives rise to the common interosseous artery?

____

63. What artery supplies the deep muscles of the posterior forearm?

____

64. What two arteries form an anastomotic connection behind the medial epicondyle?

a. ______________________________________

b. ______________________________________

65. What two arteries form an anastomotic connection in front of the medial epicondyle?

a. ______________________________________

b. ______________________________________

66. What two arteries form an anastomotic connection behind the lateral epicondyle?

a. ______________________________________

b. ______________________________________

67. What two arteries form an anastomotic connection in front of the lateral epicondyle?

a. ______________________________________

b. ______________________________________

68. Describe the fiber orientation (direction) of the interosseous membrane.

____

69. What bone is found in the floor of the “anatomical snuff box”?

____

70. What are the branches of the brachial artery not including its terminal branches?

a. ______________________________

b. ______________________________

c. ______________________________

71. What are the major terminal branches of the profunda brachial artery?

a. ______________________________

b. ______________________________

c. ______________________________

72. What nerve passes through the scapular notch (beneath the transverse scapular ligament)?

____

73. What muscle inserts on the lesser tuberosity of the humerus?

____

74. On which bone of the forearm does the tendon of the biceps brachii insert?

____

75. On which bone of the forearm does the tendon of the brachialis insert?

____

76. Which two (2) carpal bones form the radiocarpal joint?

a. ______________________________

b. ______________________________

77. What spinal nerve roots contribute to form the long thoracic nerve?

____

78. Excluding the terminal branches, what three (3) nerves are branches of the posterior cord of the brachial plexus?

a. ______________________________

b. ______________________________

c. ______________________________

79. Which of the thenar muscles is NOT innervated by the median nerve?

____

80. The ulnar nerve passes into the forearm, passing between the two heads of origin of what muscle?

____

81. The median nerve passes into the forearm, passing between the two heads of origin
of what muscle?

____

82. The radial nerve passes into the forearm, passing between the two heads of origin of what muscle?

____

4 Lower Limb

1. What two (2) bones form the pelvic girdle?

   a. ______________________________

   b. ______________________________

2. What three (3) bones form the hip bone (os coxae or innominate bone)?

   a. ______________________________

   b. ______________________________

   c. ______________________________

3. What are the two attachments of the inguinal ligament?

   a. lateral attachment ______________________________

   b. medial attachment ______________________________

4. What is the numerical value of the angle of inclination in the adult?

   ____

5. What is the numerical value of the angle of declination (torsion) in the adult?

   ____

6. What is the name of the deep fascia of the thigh?

   ____

7. What is the name of the thickened part of the deep fascia on the lateral surface of the thigh?

____

8. What is the name of the thin fascial layer that covers the saphenous opening of the fascia lata?

____

9. Into what vein does the great saphenous vein drain?

____

10. Into what vein does the small saphenous vein drain?

____

11. What spinal segments give rise to nerve fibers that form the femoral nerve?

____

12. What spinal segments give rise to nerve fibers that form the obturator nerve?

____

13. What peripheral nerve provides sensory innervation to the skin on the dorsal surface of the foot between the 1st and 2nd metatarsals?

____

14. What is the origin of the sartorius?

____

15. What is the origin of the rectus femoris?

____

16. What is the insertion of the psoas major?

____

17. The adductor hiatus is an opening in the aponeurotic attachment of what muscle?

____

18. What structure passes through the adductor hiatus?

____

19. What are the borders of the femoral triangle?

    a. superior ___________________________

    b. medial ___________________________

    c. lateral ___________________________

20. List from lateral to medial, the three (3) neurovascular structures that course through the femoral triangle.

    a. ___________________________

    b. ___________________________

    c. ___________________________

21. Is the femoral canal located (medial / lateral) to the neurovascular structures of the femoral triangle?

____

22. What is located in the femoral canal?

____

23. Which of the structures passing deep to the inguinal ligament is NOT enclosed in the femoral sheath?

____

24. What is the terminal branch of the femoral nerve?

____

25. In addition to the femoral nerve, what other nerve passes deep to the inguinal ligament to reach the thigh?

____

26. What nerve provides motor innervation to the adductor muscles of the thigh?

____

27. What is the major artery that supplies the structures of the thigh?

____

28. What is the insertion of the gluteus medius?

____

29. What three (3) muscles insert as the pes anserinus?

a. __________________________________

b. __________________________________

c. __________________________________

30. Name the nerve that innervates each of the two heads of the biceps femoris.

a. long head __________________________

b. short head __________________________

31. The sciatic nerve passes through the gluteal region by passing either under, over or through what muscle?

____

32. What nerve lies in the floor of the popliteal fossa?

____

33. Name the cutaneous nerve of the posterior surface of the leg.

____

34. What nerve lies immediately posterior to the head of the fibula?

____

35. Name the four (4) muscles of the anterior compartment of the leg.

a. ____

b. ____

c. ____

d. ____

36. What nerve provides motor innervation to the muscles of the anterior compartment of the leg?

____

37. Name the three (3) muscles of the superficial layer of the posterior compartment of the leg.

a. ____

b. ____

c. ____

38. Name the four (4) muscles of the deep layer of the posterior compartment of the leg.

a. ______________________________

b. ______________________________

c. ______________________________

d. ______________________________

39. The pulse of what artery can be palpated on the dorsal surface of the foot?

____

40. What are the two (2) actions at the ankle of the muscles of the lateral compartment of the leg?

a. ______________________________

b. ______________________________

41. What nerve provides sensory innervation to the skin over the lateral surface of the 5th metatarsal?

____

42. Which two (2) muscles of the leg exert an action on the knee joint?

a. ______________________________

b. ______________________________

43. What is the insertion of the gastrocnemius?

____

44. What nerve provides sensory innervation to the skin over the anterior compartment of the leg?

____

45. What nerve of the leg terminates as the medial and lateral plantar nerves of the foot?

____

46. Name the three (3) ligaments that reinforce the stability of the hip joint.

a. ____________________________________

b. ____________________________________

c. ____________________________________

47. What artery gives rise to the artery of the head of the femur?

____

48. What artery is the usual origin of the anterior and posterior femoral circumflex arteries?

____

49. What ligament of the knee resists posterior displacement of the tibia on the femur?

____

50. What ligament of the ankle joint is commonly referred to as the deltoid ligament?

____

51. The pulse of what artery can be palpated on the posterior to the medial malleolus?

____

52. What are the six (6) main branches of the femoral artery in the femoral triangle?

a. ______________________________

b. ______________________________

c. ______________________________

d. ______________________________

e. ______________________________

f. ______________________________

4 Thoracic Wall

1. Name the three (3) parts of the sternum.

   a. ______________________

   b. ______________________

   c. ______________________

2. Which rib attaches anteriorly at the level of the sternal angle (angle of Louis)?

   ____

3. Which ribs are defined as vertebra-sternal ribs?

   ____

4. At rest, what vertebral body would be crossed by a horizontal line extending posteriorly from the xiphisternal joint?

   ____

5. Which spinal nerve provides sensory innervation to the skin of the nipple?

   ____

6. In the midaxillary line, the intercostal vessels and nerves are located between which two intercostal muscles?

   a. ______________________

   b. ______________________

7. Name the artery that is the origin of each of the arteries listed below.

Artery	Origin
a. anterior intercostal art.	______________
b. posterior intercostal art.	______________

8. Which set of posterior intercostal arteries (left or right) passes laterally, across the vertebral body to reach the intercostal space?

____

9. The intercostal vessels and nerve lie in the costal groove. What is the order of these structures from superior to inferior?

a. ______________________________

b. ______________________________

c. ______________________________

10. What is the action of each of the muscles listed below?

Muscle	Action
a. serratus posterior inferior	______________
b. serratus posterior superior	______________

11. The female breast receives its arterial blood supply by way of the medial and lateral mammary arteries. What is the origin of each of these two sets of vessels?

Artery	Origin
a. lateral mammary art	______________
b. medial mammary art	______________

12. What are the two (2) terminal branches of the internal thoracic artery?

a. ______________________________

b. ______________________________

13. At rest, what vertebral body would be crossed by a horizontal line extending posteriorly from the sternal angle?

____

14. Which costal cartilage marks the point where the internal thoracic artery divides into its two (2) terminal branches?

____

5 Pleura and Lungs

1. Name the four (4) parts of the parietal pleura.

   a. ______________________________

   b. ______________________________

   c. ______________________________

   d. ______________________________

2. What structure passes over the top of the hilum of the lung on the right side?

   ____

3. What structure passes over the top of the hilum of the lung on the left side?

   ____

4. What is the name of the pleural structure that extends inferiorly from the root of the lung to the diaphragm?

   ____

5. At what vertebral level does the trachea divide into main stem bronchi?

   ____

6. Into what vessel do each of the following veins drain into?

	Vein	Receiving Vein
a.	left bronchial vein	______________
b.	right bronchial vein	______________

7. Which main stem bronchus is wider?

____

8. Indicate the rib overlying the pleural reflection in the:

   a. midclavicular line ______________________________

   b. midaxillary line ______________________________

   c. midscapular line ______________________________

9. Which costal cartilage marks the point where the pleural reflection moves laterally beneath the body of the sternum on the right side?

____

10. Which costal cartilage marks the point where the pleural reflection moves laterally beneath the body of the sternum on the right side?

____

11. Which rib overlies the lung boundary in the:

   a. midclavicular line ______________________________

   b. midaxillary line ______________________________

   c. midscapular line ______________________________

12. Which costal cartilage marks the point where the medial edge of the right lung moves laterally beneath the body of the sternum?

____

13. Which costal cartilage marks the point where the medial edge of the left lung moves laterally beneath the body of the sternum?

____

14. The oblique fissure lies parallel to a line connecting the:

a. ____________________ spinous process posteriorly

b. ____________________ rib in the midaxillary line

c. ____________________ costal cartilage anteriorly

15. The horizontal fissure lies parallel to a line interconnecting the:

a. ____________________ in the midaxillary line laterally

b. ____________________ costal cartilage anteriorly

16. What spinal cord segments give rise to the phrenic nerve?

____

17. Which lobe of the lung is aerated by the eparterial bronchus?

____

18. In the root of the neck, the phrenic nerve lies on the anterior surface of what muscle?

____

19. The phrenic nerve lies (anterior / posterior) to the root of the lung as it descends toward the diaphragm?

____

20. What vessels accompany the phrenic nerve as it passes the root of the lung?

____

6 Mediastinum

1. The boundary between the superior mediastinum and the inferior mediastinum can be described as a horizontal line extending from the:

   a. ______________________________ anteriorly

   b. ______________________________ posteriorly

2. Name the four (4) major branches of the left coronary artery.

   a. ______________________________

   b. ______________________________

   c. ______________________________

   d. ______________________________

3. Name the four (4) major branches of the right coronary artery.

   a. ______________________________

   b. ______________________________

   c. ______________________________

   d. ______________________________

4. Which cardiac valve is located between the right atrium and right ventricle?

   ____

5. Indicate the location of the nerve cells that provide autonomic innervation to the heart.

   a. preganglionic sympathetic ____________________

   b. postganglionic sympathetic ____________________

   c. preganglionic parasympathetic ____________________

   d. postganglionic parasympathetic ____________________

6. Name in order from proximal to distal, the three (3) arteries that arise from the arch of the aorta.

   a. ____

   b. ____

   c. ____

7. The azygos vein drains into what vein?

   ____

8. Pectinate muscles are a feature of which chamber of the heart?

   ____

9. Which artery passes horizontally, immediately posterior to the ascending aorta (be specific)?

   ____

10. Which cardiac veins do NOT drain into the coronary sinus?

   ____

11. In the superior mediastinum, what structure lies immediately posterior to the trachea?

   ____

12. The thoracic duct drains into the angle formed by which two vessels?

a. ______________________________

b. ______________________________

13. In the superior mediastinum, the left phrenic nerve lies (anterior / posterior) to the left pulmonary artery.

____

14. In the superior mediastinum, the left brachiocephalic vein lies (anterior / posterior) to the left common carotid artery.

____

15. In the superior mediastinum, the ascending aorta lies (anterior / posterior) to the trachea.

____

16. In the superior mediastinum, the esophagus lies (anterior / posterior) to the trachea.

____

17. In the superior mediastinum, the thoracic duct lies (anterior / posterior) to the superior vena cava.

____

18. In the superior mediastinum, the descending aorta lies (anterior / posterior) to the left main stem bronchus.

____

19. In the superior mediastinum, the descending aorta lies (anterior / posterior) to the left pulmonary artery.

____

20. In the posterior mediastinum, the thoracic duct lies (to the right / to the left) of the thoracic aorta.

____

21. In the posterior mediastinum, the esophagus lies (anterior / posterior) to the descending aorta.

____

22. What is the origin of the internal thoracic artery?

____

23. Name the opening in the respiratory diaphragm does NOT open into the posterior mediastinum?

____

24. The phrenic nerve courses inferiorly through the mediastinum, passing

_________________ (anterior or posterior) to the root of the lung.

25. The vagus nerve courses inferiorly through the mediastinum, passing

_________________ (anterior or posterior) to the root of the lung.

26. What vertebral level marks each of the following openings in the diaphragm?

	Opening	Vertebral level
a.	aortic hiatus	______________
b.	vena caval hiatus	______________
c.	esophageal hiatus	______________

27. Where is the best place on the chest to auscultate the aortic valve?

____

28. Where is the best place on the chest to auscultate the tricuspid valve?

____

29. Where is the best place on the chest to auscultate the pulmonary valve?

____

30. Where is the best place on the chest to auscultate the mitral valve?

____

31. Which covering of the heart is also known as the epicardium?

____

32. Which chamber forms the right border of the heart?

____

33. Which coronary artery gives rise to a branch that supplies the SA node?

____

34. What structure lies immediately behind the esophagus at the T10 vertebral level?

___

35. What structure lies on the anterior surface of the left main stem bronchus?

___

7 Abdomen

1. Name the layer of fascia located immediately deep to the transversus abdominus muscle.

____

2. What structure forms the posterior wall of the rectus sheath above the arcuate line?

____

3. Name two (2) arteries that perfuse the anterior abdominal wall, that originate from the external iliac artery?

   a. ____________________________________

   b. ____________________________________

4. Name two (2) arteries that perfuse the anterior abdominal wall that originate from the femoral artery?

   a. ____________________________________

   b. ____________________________________

5. What artery forms an anastomotic connection with the inferior epigastric artery?

____

6. What spinal nerve provides sensory innervation to the skin of the umbilicus?

____

7. What artery supplies the rectus abdominus muscle at the level of the arcuate line?

____

8. Name the attachments of the inguinal ligament.

a. superolateral ______________________

b. inferomedial ______________________

9. What structure forms the deep inguinal ring?

____

10. What structure forms the superficial inguinal ring?

____

11. The femoral nerve, artery and vein enter the thigh by coursing deep to the inguinal ligament. What is the anatomical relationship of these three structures, from medial to lateral?

a. ______________________________________

b. ______________________________________

c. ______________________________________

12. What structure contributes to each of the three (3) layers of the spermatic cord?

a. external spermatic fascia ______________________

b. cremaster fascia ______________________

c. internal spermatic fascia ______________________

13. Which of the abdominal muscles gives rise to the cremaster muscle?

____

14. What nerve provides motor innervation to the cremaster muscle?

____

15. Name the tubular structure associated with the testes where sperm cells are stored
until ejaculation.

____

16. What structure of the abdomen gives rise to the tunica vaginalis of the testes?

____

17. Into what veins do the right and left testicular veins drain?

a. right testicular vein _______________________

b. left testicular vein _______________________

18. Name the three (3) parts of the greater omentum.

a. _______________________________________

b. _______________________________________

c. _______________________________________

19. Name the two (2) parts of the lesser omentum.

a. _______________________________________

b. _______________________________________

20. What three (3) structures are located within the hepatoduodenal ligament?

a. _______________________________________

b. _______________________________________

c. _______________________________________

21. The muscular wall of the esophagus is formed by both smooth and striated muscle. Indicate the type of muscle that is found in each of the three (3) regions of the esophagus listed below.

a. proximal 1/3 ____________________

b. middle 1/3 ____________________

c. distal 1/3 ____________________

22. What nerve provides motor innervation to each of the three (3) regions of the esophagus?

a. proximal 1/3 ____________________

b. middle 1/3 ____________________

c. distal 1/3 ____________________

23. Name the four (4) parts (regions) of the stomach.

a. ____________________

b. ____________________

c. ____________________

d. ____________________

24. What arterial branch of the aorta gives rise to branches that supply the stomach?

____

25. What arteries form the anastomotic network located on the lesser curvature of the stomach AND what is the origin of each of these arteries?

	Artery	Origin
a.	____________	____________
b.	____________	____________

26. What arteries form the anastomotic network located on the greater curvature of the stomach AND what is the origin of each of these arteries?

	Artery	Origin
a.	____________________	____________________
b.	____________________	____________________

27. Name the three (3) major branches of the celiac trunk.

a. ____

b. ____

c. ____

28. Name the three (3) branches of the common hepatic artery.

a. ____

b. ____

c. ____

29. Name the two (2) branches of the gastroduodenal artery.

a. ____

b. ____

30. Name the two (2) veins that drain into the portal vein.

a. ____

b. ____

31. What two (2) structures form the common bile duct?

a. ____

b. ____

32. What two (2) structures drain into the hepatopancreatic ampulla (ampulla of Vater)?

a. ______________________________

b. ______________________________

33. Into which part of the duodenum does the hepatopancreatic ampulla drain?

____

34. What structure marks the point where the hepatoduodenal papilla enters the duodenum?

____

35. What structure drains into the duodenum by way of the minor duodenal papilla?

____

36. At what level of the vertebral column does the horizontal part (3rd part) of the duodenum cross the vertebral column?

____

37. What artery lies immediately anterior to the horizontal part of the duodenum?

____

38. What structure "supports/suspends" the small intestine at the level of the duodenal-jejunal junction?

____

39. Which parts of the duodenum are considered to be retroperitoneal?

____

40. Which parts of the duodenum are supplied by branches of the celiac artery?

____

41. The superior mesenteric artery supplies arterial blood to the intestine. Indicate the proximal most part and the distal most part of the intestine that receives arterial blood supply by way of branches of the superior mesenteric artery.

    a. proximal most part ________________________

    b. distal most part ________________________

42. What part(s) of the intestine receive arterial blood supply by way of branches of the inferior mesenteric artery?

____

43. What part(s) of the intestine are drained by branches of the inferior mesenteric vein?

____

44. Which parts of the colon are retroperitoneal?

____

45. What three (3) structures form the portal triad?

    a. ____________________________________

    b. ____________________________________

    c. ____________________________________

46. What structure divides the liver into left and right lobes?

____

47. Which two (2) lobes of the liver lie on either side of the porta hepatis?

a. ______________________________

b. ______________________________

48. The round ligament (ligamentum teres) of the liver is a remnant of what embryological structure?

____

49. The ligamentum venosum of the liver is the remnant of what embryological structure?

____

50. In its course to the right kidney, the right renal artery passes (anterior / posterior) to the inferior vena cava.

____

51. In its course from the left kidney to the inferior vena cava, the left renal vein passes (anterior / posterior) to the aorta.

____

52. Indicate the origin of each of the suprarenal arteries listed below.

	Suprarenal art.	Origin
a.	superior suprarenal art.	____________
b.	middle suprarenal art.	____________
c.	inferior suprarenal art.	____________

53. Which spinal cord segments are the origin of the greater splanchnic nerve?

____

54. Which spinal cord segments are the origin of the lesser splanchnic nerve?

____

55. Name the structure that forms the lateral umbilical fold.

____

56. Name the structure that forms the medial umbilical fold.

____

57. At what vertebral level does the aorta bifurcate to form the common iliac arteries?

____

58. At what vertebral level do the common iliac veins join to form the inferior vena cava?

____

59. What are the five (5) branches of the superior mesenteric artery?

a. ____________________________________

b. ____________________________________

c. ____________________________________

d. ____________________________________

e. ____________________________________

60. What are the three (3) branches of the inferior mesenteric artery?

a. ____________________

b. ____________________

c. ____________________

61. What paired lymphatic structures drain into the cisterna chyli?

____

62. What two (2) nerves lie in the inguinal canal?

a. ____________________

b. ____________________

63. Are the tendinous intersections of the abdominal wall located above or below the arcuate line?

____

64. Name the plexus of nerves that lies on the anterior surface of the bifurcation of the aorta.

____

65. What section of the small intestine is associated with Meckel's diverticulum?

____

66. What are the three (3) parts of the teniae coli?

a. ____________________

b. ____________________

c. ____________________

67. What nerve lies on the anterior surface of the psoas major muscle?

____

68. What are the two (2) components of the superficial layer of abdominal fascia?

a. ______________________________

b. ______________________________

69. What component of the spermatic cord is formed by the deep fascia of the abdomen?

____

70. What component of the penis is formed by the deep fascia of the abdomen?

____

71. What are the eight (8) branches of the lumbar plexus?

a. ______________________________

b. ______________________________

c. ______________________________

d. ______________________________

e. ______________________________

f. ______________________________

g. ______________________________

h. ______________________________

72. Name the tributaries of the inferior vena cava.

a. ______________________________

b. ______________________________

c. ______________________________

d. ______________________________

e. ______________________________

f. ______________________________

73. Name the origin of each of the arteries listed below that perfuse the stomach.

	Gastric Vessel	Origin
a.	short gastric artery	______________
b.	left gastric artery	______________
c.	right gastric artery	______________
d.	left gastroepiploic artery	______________
e.	right gastroepiploic artery	______________

74. Name the muscles hat pass beneath the medial and lateral arcuate ligaments of the respiratory diaphragm.

a.	medial arcuate ligament	______________
b.	lateral arcuate ligament	______________

75. Name the three (3) abdominal organs with veins that drain into the splenic vein.

a. ______________________________

b. ______________________________

c. ______________________________

76. Name the three (3) borders of the lumbar triangle.

a. ______________________________

b. ______________________________

c. ______________________________

77. Name the three (3) borders of the inguinal triangle.

a. ______________________________

b. ______________________________

c. ______________________________

78. What embryonic structure is located in the free margin of the falciform ligament?

____

8 Pelvis amd Perineum

1. Name the three (3) bones that form the "hip" bone (coxal bone).

   a. ______________________________

   b. ______________________________

   c. ______________________________

2. Which part of the pelvis is located above (superior to) the pelvic brim?

   ____

3. What ligament converts the sciatic notch into the sciatic foramen?

   ____

4. What ligament subdivides the sciatic foramen into the greater and lesser sciatic foraminae?

   ____

5. What two (2) muscles form the pelvic diaphragm?

   a. ______________________________

   b. ______________________________

6. What three (3) muscles form the levator ani?

   a. ______________________________

   b. ______________________________

   c. ______________________________

7. What muscle lies in (passes through) the greater sciatic foramen?

____

8. What muscle forms the lateral border of the urogenital hiatus?

____

9. Name the peritoneal fold that, in the female, extends from the uterus to the lateral wall of the pelvis

____

10. What muscle lies between the superior and inferior pelvic fascia?

____

11. What is the origin of the obturator artery?

____

12. What is the origin of the internal pudendal artery?

____

13. Into what vein does the internal iliac vein drain?

____

14. What nerve is the main nerve of the perineum and the main sensory nerve of the external genitalia?

____

15. What spinal segments give rise to parasympathetic fibers that innervate the structures of the pelvis?

____

16. What spinal segments contribute to the formation of the pudendal nerve?

____

17. What nervous system structure interconnects the superior hypogastric plexus with the inferior hypogastric plexus?

____

18. Do the ureters course (anterior or posterior) to the common iliac artery as they enter the pelvis?

____

19. What is the name of the muscle that forms the wall of the urinary bladder?

____

20. What type of muscle forms the internal urethral sphincter?

____

21. What type of muscle forms the external urethral sphincter?

____

22. Name the nerves formed by parasympathetic fibers that provide motor innervation to the urinary bladder.

____

23. What structures mark the superior and inferior angles of the trigone of the urinary bladder?

    a. superior angles ____________________________

    b. inferior angle ____________________________

24. Indicate the origin of the three (3) arteries that supply the rectum

a. superior rectal art. ____________________

b. middle rectal art. ____________________

c. inferior rectal art. ____________________

25. What two (2) structures drain into the ejaculatory duct?

a. ____________________

b. ____________________

26. What glands are embedded in the external urethral sphincter?

____

27. Name the four (4) parts of the uterine tube in order from distal to proximal

a. ____________________

b. ____________________

c. ____________________

d. ____________________

28. What is the name of the large mesentery that extends from the sides of the uterus to the lateral wall and floor of the pelvis?

____

29. What are the two (2) main parts of the uterus?

a. ____________________

b. ____________________

30. Into what veins do the right and left ovarian veins commonly drain?

a. right ovarian vein ______________________________

b. left ovarian vein ______________________________

31. Into what veins do the right and left testicular veins drain?

a. right testicular vein ______________________________

b. left testicular vein ______________________________

32. Name the three (3) layers of the uterine wall from deep to superficial.

a. ____

b. ____

c. ____

33. Name the space formed where the peritoneum "folds" from the urinary bladder to the uterus.

____

34. Name the space formed where the peritoneum "folds" from the rectum to the uterus.

____

35. The perineum is divided into two (2) triangles by a line passing between the right and left ischial tuberosities. What are the names of the two triangles formed by this line?

a. anterior triangle ______________________________

b. posterior triangle ______________________________

36. What are the boundaries of the deep perineal pouch (space)?

a. superior boundary ______________________

b. inferior boundary ______________________

37. In the male, what muscle is located in the deep perineal pouch (space)?

____

38. In the male, what three (3) muscles are located in the superficial perineal pouch (space)?

a. ______________________________________

b. ______________________________________

c. ______________________________________

39. The pudendal canal (Alcock's canal) is a small neurovascular passageway through which pass the pudendal vessels and nerve. This canal lies on the medial surface of what muscle of the pelvis?

____

40. What type of muscle forms the internal anal sphincter?

____

41. What type of muscle forms the external anal sphincter?

____

42. What anatomical landmark marks the point of anastomosis between blood supply to the anus delivered by the superior rectal artery and the inferior rectal artery?

____

43. What anatomical landmark marks point of transition between the part of the anus with epithelium derived from endoderm (hindgut) and the part with epithelium derived from ectoderm (proctodeum)?

____

44. What is the name of the single (unpaired) cylindrical structure forming part of the shaft of the penis?

____

45. What is the name of the paired cylindrical structures forming part of the shaft of the penis?

____

46. The skin of the scrotum is innervated by two nerves: the anterior and posterior scrotal nerves. What two (2) nerves contribute to each of these scrotal nerves?

anterior scrotal nerve

a. ____________________________________

b. ____________________________________

posterior scrotal nerve

c. ____________________________________

d. ____________________________________

47. What artery is the origin of branches that supply corpora of the penis?

____

48. What glands are located on either side of the vestibule of the vagina?

____

49. Name the peritoneal fold that, in the female, extends from the ovary to the fallopian tube.

____

50. Not including the femoral artery, what are the two (2) main branches of the external
iliac artery?

a. ________________________________

b. ________________________________

51. What are the eleven (11) branches of the internal iliac artery?

a. ________________________________

b. ________________________________

c. ________________________________

d. ________________________________

e. ________________________________

f. ________________________________

g. ________________________________

h. ________________________________

i. ________________________________

j. (female) ________________________

k. (female) ________________________

52. Into what vein do each of the three (3) rectal veins drain?

a. ________________________________

b. ________________________________

c. ________________________________

53. What are the three (4) parts of the male urethra?

a. ______________________________

b. ______________________________

c. ______________________________

d. ______________________________

54. What is the thick external covering of the testes?

____

55. What is the dense plexus of veins on the surface of the testes that drains into the testicular vein?

____

56. What structure connects the epididymis to the ejaculatory duct?

____

57. Wat two (2) muscles lie in the deep perineal space in the male?

a. ______________________________

b. ______________________________

58. What nerve innervates the muscles in the deep perineal space?

____

59. What structure in the female is the homologue of the male bulbourethral glands?

____

60. What nerve provides motor innervation to the external anal sphincter?

____

9 Head and Face

1. What bones are joined to form the lambda?

   a. ______________________________

   b. ______________________________

2. What bones are joined to form the bregma?

   a. ______________________________

   b. ______________________________

3. What two (2) bones contribute to the zygomatic arch?

   a. ______________________________

   b. ______________________________

4. What four (4) bones contribute to the formation of the pterion?

   a. ______________________________

   b. ______________________________

   c. ______________________________

   d. ______________________________

5. At approximately what age does the anterior fontanelle become no longer palpable?

   ____

6. What structure divides the intracranial compartment into right and left halves?

   ____

7. What structure divides the intracranial compartment into supratentorial and infratentorial compartments?

____

8. What are the five (5) layers of the scalp?

a. ______________________________________

b. ______________________________________

c. ______________________________________

d. ______________________________________

e. ______________________________________

9. What branch of the external carotid artery:

a. is palpable along the inferior border of the mandible?

____

b. supplies the structure of the tongue?

____

c. is palpable in the temporal fossa?

____

d. supplies part of the thyroid gland?

____

e. enters the pterygopalatine fossa?

____

10. What artery anastomoses from the front with the intraorbital branches of the ophthalmic artery?

____

11. What dural venous sinus:

a. lies in the superior margin of the falx cerebri?

____

b. lies in the inferior margin of the falx cerebri?

____

c. lies along the petrous ridge?

____

d. lies immediately lateral to the sella turcica?

____

e. at the junction of the falx cerebri and the tentorium cerebelli?

____

12. Which two (2) dural venous sinuses drain directly into the jugular vein?

a. ______________________________

b. ______________________________

13. What artery enters the cranial cavity by passing through the foramen magnum?

____

14. What artery enters the cranial cavity by passing through the foramen spinosum?

____

15. What two (2) arteries are connected by way of the anterior communication artery?

a. ______________________________

b. ______________________________

16. What two (2) arteries are connected by way of the posterior communication artery?

a. ______________________________

b. ______________________________

17. What are the five (5) major intracranial branches of the internal carotid artery?

a. ______________________________

b. ______________________________

c. ______________________________

d. ______________________________

e. ______________________________

18. What cranial nerve passes through the parotid gland?

____

19. What are the four (4) muscles of mastication?

a. ______________________________

b. ______________________________

c. ______________________________

d. ______________________________

20. What cranial nerve mediates sensation from the:

a. forehead over the eyebrows

____

b. skin over the maxilla?

____

c. skin over the mental tubercle

____

21. What cranial nerve provides parasympathetic innervation to the parotid gland?

____

a. What is the location of the postganglionic cell body?

____

22. What cranial nerve provides parasympathetic innervation to the submandibular and submaxillary glands?

____

a. What is the location of the postganglionic cell body?

____

23. What cranial nerve provides parasympathetic innervation to the lacrimal gland?

____

a. What is the location of the postganglionic cell body?

____

24. What cranial nerve provides motor innervation to the muscles of facial expression?

____

25. What cranial nerve provides motor innervation to the muscles of mastication?

____

26. What cranial nerve provides sensory innervation to the face?

____

a. What branch of this nerve innervates the skin over the eyebrow?

____

1) Which opening in the middle cranial fossa contains the axons of this nerve branch?

____

b. What branch of this nerve innervates the skin over the maxilla?

____

1) Which opening in the middle cranial fossa contains the axons of this nerve branch?

____

c. What branch of this nerve innervates the skin over the mental protuberance?

____

1) Which opening in the middle cranial fossa contains the axons of this nerve branch?

____

27. What foramen transmits the axons of the glossopharyngeal nerve?

____

28. What striated (skeletal) muscle is innervated by the glossopharyngeal nerve?

____

29. What special sensory functions are mediated by the glossopharyngeal nerve?

a. ______________________________

b. ______________________________

30. What foramen transmits the axons of the vagus nerve?

____

31. What striated muscles are innervated by the vagus nerve?

a. ______________________________

b. ______________________________

32. What is the effect of the vagus nerve on heart rate?

____

33. What is the effect of the vagus nerve on gastric and intestinal motility?

____

34. What special sensory functions are mediated by the vagus nerve?

a. ______________________________

b. ______________________________

35. What two foramina transmit the axons of the spinal accessory nerve?

a. ______________________________

b. ______________________________

36. What muscles are innervated by the spinal accessory nerve?

a. ______________________________

b. ______________________________

37. What foramen transmits the axons of the hypoglossal nerve?

____

38. What muscle innervated by the hypoglossal nerve is primarily involved in protrusion of the tongue?

____

39. What cranial nerve exits the skull via the stylomastoid foramen?

____

40. Through what foramen does the facial nerve enter the skull (exit the posterior cranial fossa)?

____

41. Through what foramen does the facial nerve exit the skull?

____

42. What are the five terminal branches of the facial nerve that emerge from the substance of the parotid gland?

a. ______________________________

b. ______________________________

c. ______________________________

d. ______________________________

e. ______________________________

43. What two (2) nerves form the nerve of the pterygoid canal (Vidian nerve)?

a. ______________________________

b. ______________________________

44. What cranial nerve gives rise to the chorda tympani?

____

45. What cranial nerve gives rise to the greater superficial petrosal nerve?

____

46. What cranial nerve gives rise to the tympanic nerve?

____

47. What two (2) nerves provide sensory innervation to the meninges surrounding the brain?

a. ______________________________

b. ______________________________

48. Other than the muscles of mastication and the muscles of the scalp, what four (4) skeletal muscles are innervated by the trigeminal nerve?

a. ______________________________

b. ______________________________

c. ______________________________

d. ______________________________

49. Other than the muscles of facial expression and the auricular muscles, what three (3) skeletal muscles are innervated by the facial nerve?

a. ______________________________

b. ______________________________

c. ______________________________

50. Which intracranial foramina transmits the axons of the ophthalmic nerve?

____

51. Which intracranial foramina transmits the axons of the maxillary nerve?

____

52. Which intracranial foramina transmits the axons of the mandibular nerve?

____

53. Which intracranial foramina transmits the axons of the optic nerve?

____

54. What three (3) cranial nerves pass through the internal auditory (acoustic) meatus?

a. ______________________________

b. ______________________________

c. ______________________________

55. What nerve passes through the mandibular foramen?

____

56. What autonomic ganglion is located in the orbit?

____

57. What autonomic ganglion is located in the pterygopalatine fossa?

____

58. What autonomic ganglion is located immediately below the foramen ovale?

____

59. What autonomic ganglion is located in the floor of the mouth?

____

60. What nerve passes from the infratemporal fossa to the orbit by way of the inferior orbital fissure?

____

61. What artery lies in the cavernous sinus?

____

62. What nerve passes through the cavernous sinus?

____

63. What nerve passes through the sphenopalatine foramen?

____

64. What artery passes into the pterygopalatine fossa through the pterygomaxillary fissure?

____

65. What nerve exits the skull by way of th stylomastoid foramen?

____

66. Which cranial nerve gives rise to parasympathetic fibers that are part of the lesser petrosal nerve?

____

67. Which cranial nerve gives rise to parasympathetic fibers that are part of the Tympanic nerve?

____

68. Where are the postganglionic parasympathetic nerve cell bodies that receive synaptic input from the fibers of the lesser petrosal nerve?

____

69. What nerve emerges from the mental foramen?

____

70. The glabella is a feature of what bone?

____

71. The cribriform plate is a feature of what bone?

____

72. The sella turcica is a feature of what bone?

____

73. The styloid process is a feature of what bone?

____

74. What two (2) bones form the clivus?

a. ______________________________

b. ______________________________

75. What veins transmit blood from the scalp to the dural venous sinuses?

____

76. What veins are located between the inner and outer layers of the calvarium?

____

77. What two (2) venous structures merge to form the straight sinus?

____

78. What nerve exits the skull by way of the petrotympanic fissure?

____

10 Mouth and Pharynx

1. What are the two (2) main muscles are used to close the mouth?

   a. ____________________

   b. ____________________

2. What muscle forms the floor of the oral cavity?

   ____

3. What muscle forms that palatoglossal arch?

   ____

   a. What nerve provides motor innervation to this muscle?

      ____

4. What muscle forms the palatopharyngeal arch?

   ____

   a. What nerve provides motor innervation to this muscle?

      ____

5. What muscle forms the anterior pillar of the tonsillar fossa?

   ____

6. What muscle forms the posterior pillar of the tonsillar fossa?

   ____

7. What nerve provides sensory innervation to the mucosal lining of the tonsillar fossa?

____

8. What two (2) muscles act on the soft palate?

a. ____

b. ____

9. What nerve provides motor innervation to the superior, middle and inferior pharyngeal constrictor muscles?

____

10. What nerve provides motor innervation to the intrinsic muscles of the tongue?

____

11. What nerve provides motor innervation to MOST of the extrinsic muscles of the tongue?

____

12. Which of the extrinsic muscles of the tongue is NOT innervated like the others?

____

a. What nerve provides motor innervation to this muscle?

____

13. What nerve innervates mechanical and thermal receptors on the anterior 2/3 of the tongue?

____

14. What nerve innervates mechanical and thermal receptors on the posterior 1/3 of the tongue?

____

15. What nerve innervates taste buds on the anterior 2/3 of the tongue?

____

16. What nerve innervates taste buds on the posterior 1/3 of the tongue?

____

17. What cranial nerve provides motor innervation to the submandibular salivary glands?

____

18. What cranial nerve provides motor innervation to the parotid salivary glands?

____

19. What nerve provides sensory innervation to the gingiva of the maxillary teeth?

____

20. What nerve provides sensory innervation to the gingiva of the mandibular teeth?

____

21. What nerve provides sensory innervation to the mucosa overlying the hard palate?

____

22. What nerve provides sensory innervation to the mucosa overlying the soft palate?

____

23. What feature of the tongue marks the attachment of the thyroglossal duct?

____

24. What nerve provides motor innervation to each of the following muscles of the pharynx?

a. palatopharyngeus _______________________

b. salpingopharyngeus _______________________

c. stylopharyngeus _______________________

25. What structure lies between the palatoglossal and palatopharyngeal folds?

____

26. What nerve innervates the buccinator muscle?

____

27. What are the origins of each of the three (3) pharyngeal constrictors?

a. superior constrictor _______________________

b. middle constrictor _______________________

c. inferior constrictor _______________________

28. Name the three (3) pharyngeal muscles other that the constrictor muscles.

a. _______________________________________

b. _______________________________________

c. _______________________________________

29. What tubular structure connects the nasopharynx with the tympanic cavity?

____

30. Name the potential space located between the buccopharyngeal fascia and the prevertebral fascia.

____

11 Neck and Larynx

1. What are the four (4) parts of the deep cervical fascia?

   a. ______________________________

   b. ______________________________

   c. ______________________________

   d. ______________________________

2. What two (2) muscles are ensheathed by the investing layers of the deep cervical fascia?

   a. ______________________________

   b. ______________________________

3. What four (4) structures are enclosed within the pretracheal layer of deep cervical fascia?

   a. ______________________________

   b. ______________________________

   c. ______________________________

   d. ______________________________

4. What are the three (3) main structures enclosed within the carotid sheath?

   a. ______________________________

   b. ______________________________

   c. ______________________________

5. What are the boundaries of the anterior triangle of the neck?

a. ______________________________

b. ______________________________

c. ______________________________

6. What are the boundaries of the carotid triangle?

a. ______________________________

b. ______________________________

c. ______________________________

7. What are the boundaries of the submandibular triangle?

a. ______________________________

b. ______________________________

c. ______________________________

8. What are the boundaries of the submental triangle?

a. ______________________________

b. ______________________________

c. ______________________________

9. What muscles are boundaries of the muscular triangle?

a. ______________________________

b. ______________________________

c. ______________________________

10. What are the boundaries of the posterior triangle of the neck?

a. ______________________________

b. ______________________________

c. ______________________________

11. What muscles comprise the infrahyoid muscles of the neck?

a. ______________________________

b. ______________________________

c. ______________________________

d. ______________________________

12. What muscles comprise the suprahyoid muscles of the neck?

a. ______________________________

b. ______________________________

c. ______________________________

d. ______________________________

13. What muscle abducts the vocal folds?

____

14. What nerve provides motor innervation to this muscle?

____

15. What two muscles adduct the vocal folds?

a. ______________________________

b. ______________________________

16. What nerve provides motor innervation to these muscles?

____

17. What laryngeal muscle lies on the external surface of the larynx?

____

18. What nerve provides motor innervation to this muscle?

____

19. What nerve provides sensory innervation to the mucosal lining the internal surface of the larynx?

____

20. The roots of the brachial plexus course through the root of the neck by passing between which two of the scalene muscles?

a. ____________________________________

b. ____________________________________

21. The subclavian artery courses through the root of the neck by passing ____________________ to the anterior scalene muscle.

22. The subclavian vein courses through the root of the neck by passing ____________________ to the anterior scalene muscle.

23. What is the anatomical landmark that marks the point where the subclavian artery becomes the axillary artery?

____

24. What are the five (5) branches of the subclavian artery?

a. ______________________________

b. ______________________________

c. ______________________________

d. ______________________________

e. ______________________________

25. What artery is the origin of the superior thyroid artery?

____

26. What artery is the origin of the inferior thyroid artery?

____

27. What veins merge to form the retromandibular vein?

a. ______________________________

b. ______________________________

28. What veins merge to form the external jugular vein?

a. ______________________________

b. ______________________________

29. What veins merge to form the brachiocephalic vein?

a. ______________________________

b. ______________________________

30. What spinal segments give rise to the phrenic nerve?

____

31. What spinal segments give rise to the ansa cervicalis?

____

32. At what vertebral level does the common carotid artery bifurcate to form the internal and external carotid arteries?

____

33. Sympathetic chain ganglia from what segmental levels fuse to form the superior cervical ganglion?

____

34. Sympathetic chain ganglia from what segmental levels fuse to form the middle cervical ganglion?

____

35. Sympathetic chain ganglia from what segmental levels fuse to form the inferior cervical ganglion?

____

36. Name the six (6) branches of the external carotid artery (not including the two terminal branches).

a. ____________________________

b. ____________________________

c. ____________________________

d. ____________________________

e. ____________________________

f. ____________________________

37. Name the two (2) terminal branches of the external carotid artery.

a. ______________________________

b. ______________________________

38. What are the three (3) branches of the thyrocervical trunk?

a. ______________________________

b. ______________________________

c. ______________________________

39. What are the four (4) branches of the cervical plexus?

a. ______________________________

b. ______________________________

c. ______________________________

d. ______________________________

40. What muscle divides the subclavian artery into three parts?

____

41. Into what veins do the three (3) thyroid veins drain?

a. ______________________________

b. ______________________________

c. ______________________________

42. What two (2) triangles are found within the posterior triangle of the neck?

a. ______________________________

b. ______________________________

43. What are the borders of the occipital triangle?

a. ____________________

b. ____________________

c. ____________________

44. What are the borders of the omoclavicular triangle?

a. ____________________

b. ____________________

c. ____________________

45. What muscle divides the neck into anterior and posterior triangles?

____

46. What muscle separates the subclavian artery from the subclavian vein in the root of the neck?

____

47. What nerve provides motor innervation to the sternocleidomastoid?

____

48. What is the name of the posterior primary ramus of the C2 spinal nerve?

____

49. What type of receptors are found in each of the following?

a. carotid body ____________________

b. carotid sinus ____________________

50. Name the three (3) major branches of the arch of the aorta in order from proximal to distal.

a. ____________________

b. ____________________

c. ____________________

51. What nerve in the neck lies on the anterior surface of the anterior scalene?

____

52. Beneath what structure does the recurrent laryngeal nerve "recur" on each side?

a. right side ____________________

b. left side ____________________

53. What is the origin of each of the following arteries that supply the thyroid gland?

a. inferior thyroid artery ____________________

b. superior thyroid artery ____________________

54. Into what veins do each of the three (3) thyroid veins drain?

a. superior thyroid vein ____________________

b. middle thyroid vein ____________________

c. inferior thyroid vein ____________________

55. What vertebral body would be crossed by a horizontal line passing posteriorly from superior edge of the thyroid cartilage?

____

56. Name the membrane that fills the space anteriorly between the thyroid cartilage and the cricoid cartilage.

____

57. Name the membrane that fills the space anteriorly between the hyoid bone and the thyroid cartilage.

____

58. What laryngeal muscle is NOT innervated by the inferior laryngeal nerve?

____

59. What two (2) structures pass through the thyrohyoid membrane?

 a. nerve ______________________________

 b. artery ______________________________

60. What nerve provides sensory innervation to the mucosal lining of the larynx?

____

61. Name the membrane through which the internal laryngeal nerve passes to enter the larynx.

____

62. Name the laryngeal cartilage that serves as the posterior attachment of the vocal ligament.

____

63. What ligamentous structure does the vertebral artery pass through immediately before entering the skull?

____

64. What nerve innervates the platysma muscle?

____

65. Which tracheal rings are covered by the isthmus of the thyroid gland?

____

66. What are the seven (7) tributaries of the internal jugular vein?

a. _______________________________________

b. _______________________________________

c. _______________________________________

d. _______________________________________

e. _______________________________________

f. _______________________________________

g. _______________________________________

67. What two (2) branches of the vagus nerve arise in the jugular fossa?

a. _______________________________________

b. _______________________________________

68. Which pharyngeal constrictor lies in proximity to the superior parathyroid gland?

____

69. What nerve passes through the thyrohyoid membrane?

____

70. Which branch of the external carotid artery carries postganglionic sympathetic fibers to the sublingual salivary glands?

____

71. Which triangle of the neck do the subclavian artery and vein pass through?

____

72. Through which triangle of the neck does the spinal accessory nerve pass?

____

73. Which part of the brachial plexus passes between the anterior and middle scalene muscles?

____

74. Which cervical sympathetic ganglia contributes fibers to the brachial plexus?

____

75. What six (6) cartilages comprise the larynx?

a. ________________________________

b. ________________________________

c. ________________________________

d. ________________________________

e. ________________________________

f. ________________________________

76. Name the two (2) mucosal folds that border the laryngeal ventricle.

a. ________________________________

b. ________________________________

77. What muscle form part of the laryngeal vocal fold?

____

78. Which cervical vertebra typically does not have a transverse foramen?

____

79. What nerve lies adjacent to the external jugular vein on the superficial surface of the sternocleidomastoid muscle?

____

80. What five (5) nerves emerge from the posterior edge of the sternocleidomastoid into the posterior triangle of the neck?

a. _______________________________________

b. _______________________________________

c. _______________________________________

d. _______________________________________

e. _______________________________________

81. What nerve passes between the middle and posterior scalene muscles into the posterior triangle of the neck?

____

82. What two (2) arteries course laterally on the anterior scalene muscle into the posterior triangle of the neck?

a. _______________________________________

b. _______________________________________

83. Does the omohyoid muscle lie (anterior/posterior) to the external jugular vein in the carotid triangle?

____

84. Which of the infrahyoid (strap) muscles does not have an attachment to the hyoid bone?

12 Orbit

1. What are the three (3) openings located in the posterior part of the orbit?

   a. ______________________________

   b. ______________________________

   c. ______________________________

2. What are the two (2) main structures that pass through the optic canal?

   a. ______________________________

   b. ______________________________

3. What are the main structures that pass through the superior and inferior orbital fissures?

   a. ______________________________ nerve

   b. ______________________________ nerve

   c. ______________________________ nerve

   d. ______________________________ nerve

   e. ______________________________ nerve

   f. ______________________________ nerve

   g. ______________________________ vein

   h. ______________________________ vein

4. What are the main structures that pass through the common tendinous ring?

a. ______________________________ nerve

b. ______________________________ nerve

c. ______________________________ nerve

d. ______________________________ nerve

5. What is the name of the thin membrane that lines the surface of the cornea and deep surface of the upper and lower eye lids (palpebrae)?

____

6. What two (2) muscles play a role in elevation of the upper eyelid?

a. ______________________________

b. ______________________________

7. What are the two (2) main types of glands in the upper eyelid?

a. ______________________________

b. ______________________________

8. Describe the location of the lacrimal gland in the orbit.

____

9. Where does the nasolacrimal duct drain?

____

10. What cranial nerve provides motor innervation to the lacrimal gland?

____

11. Which extraocular muscle does NOT originate from the common tendinous ring?

____

12. What nerve provides motor innervation to the levator palpebrae superioris muscle?

____

13. Where are the nerve cell bodies that provide motor innervation to Mueller's muscle?

____

14. What are the three (3) branches of the ophthalmic artery in the orbit?

a. ___________________________________

b. ___________________________________

c. ___________________________________

15. What nerve enters the orbit by way of the inferior orbital fissure?

____

16. What nerve gives rise to the short ciliary nerves?

____

17. What nerve gives rise to the long ciliary nerves?

____

18. What artery gives rise to the central artery of the retina?

____

19. Into what venous structure do the superior and inferior ophthalmic veins drain?

____

13 Eye

1. What are the three (3) layers of the eye? (List layers from superficial to deep.)

   a. ______________________________

   b. ______________________________

   c. ______________________________

2. What nerve provides sensory innervation to the cornea?

   ____

3. What structure separates the anterior chamber from the posterior chamber of the eye?

   ____

4. What fluid is found in the anterior chamber of the eye?

   ____

5. What fluid is found in the posterior chamber of the eye?

   ____

6. What two (2) muscles are located in the iris AND what division of the autonomic nervous system provides motor innervation to each?

	Muscle	Autonomic Division
a.	______________	______________
b.	______________	______________

7. What structure connects the lens to the ciliary muscle?

____

8. What is the effect of contraction of the ciliary muscle?

____

9. Where are the nerve cell bodies that provide motor innervation to the ciliary muscle?

   a. preganglionic cell _______________________

   b. postganglionic cell _______________________

10. Where are the cell bodies that provide motor innervation to the pupillary constrictors?

   a. preganglionic cell _______________________

   b. postganglionic cell _______________________

11. Where are the cell bodies that provide motor innervation to the pupillary dilators?

   a. preganglionic cell _______________________

   b. postganglionic cell _______________________

12. What structure produces aqueous humor?

____

13. Through what opening does aqueous humor normally pass to exit the eye?

____

14. Where is this opening located?

____

15. What is the optic disc?

____

16. What is the physiologic cup?

____

17. What is the macula lutea?

____

18. What is the fovea centralis?

____

19. What retinal structure is associated with the “blind spot” in the visual field?

____

20. What are the two (2) types of photoreceptor cells?

a. ______________________________________

b. ______________________________________

21. Which photoreceptor cell is heavily concentrated in the fovea centralis?

____

22. What vessel is the origin of the central artery of the retina?

____

23. The superior and inferior ophthalmic veins receive venous blood from intra-orbital structures. Blood in these vessels can drain out of the orbit by passing posteriorly, anteriorly or inferiorly, depending on intravenous pressure. What are the three (3) vascular structures that receive venous blood from the ophthalmic veins?

a. ______________________________ posteriorly

b. ______________________________ inferiorly

c. ______________________________ anteriorly

14 Ear

1. What peripheral nerves provide sensory innervation to the auricle?

   a. ______________________________

   b. ______________________________

2. Which branchial arch is the origin of each of the following ossicles?

Auditory Ossicle	Branchial Arch
Malleus	______________
Incus	______________
Stapes	______________

3. Which of the ossicles is attached to the internal surface of the tympanic membrane?

   ____

4. Which of the ossicles is attached to the oval window?

   ____

5. What muscle attaches to the malleus?

   ____

   a. What cranial nerve innervates this muscle?

   ____

6. What muscle attaches to the stapes?

____

a. What cranial nerve innervates this muscle?

____

7. What cranial nerve transmits auditory impulses to the brainstem?

____

8. What foramen of the skull transmits the axons of the vestibulocochlear nerve?

____

9. What vestibular system receptor structure is responsive to rotatory (angular) acceleration and deceleration of the head?

____

10. What vestibular system receptor structure is responsive to linear acceleration and deceleration of the head?

____

11. What fluid substance is found in the scala tympani?

____

12. What fluid substance is found in the scala vestibuli?

____

13. What fluid substance is found in the scala media?

____

14. What fluid substance is found in the semicircular canals?

____

15. Which vestibular receptor structures are associated with otoconia?

____

16. Where are the cell bodies of the afferent nerve fibers of the cochlear nerve?

____

17. Where are the cell bodies of the afferent nerve fibers of the vestibular nerve?

____

18. In which quadrant of the tympanic membrane can you see the "cone of light"?

____

19. Which two nerves pass through the tympanic cavity?

a. ______________________________________

b. ______________________________________

20. Which pharyngeal arch gives rise to the stapes?

____

21. Which pharyngeal arch gives rise to the malleus?

____

Part 2: Answer Key

1 Back and Spine

1. a. eight (8)
   b. twelve (12)
   c. five (5)
   d. five (5)

2. a. cervical spine
   b. lumbar spine

3. a. anterior longitudinal ligament
   b. ligamentum flavum
   c. interspinous ligament
   d. posterior longitudinal ligament

4. ligamentum nuchae

5. zygapophyseal (facet) joint

6. cervical vertebrae

7. C7

8. C2 (axis)

9. C7

10. a. lamina
    b. pedicle

11. a. nucleus pulposus
    b. anulus fibrosus

12. posterior atlanto-occipital membrane

13. internal vertebral plexus (Batson's plexus)

14. thoracolumbar fascia (thoracodorsal fascia, lumbodorsal fascia)

15. a. trapezius
    b. latissimus dorsi
    c. rhomboid major
16. a. splenius capitis
    b. splenius cervicis

17. erector spinae muscles

18. transversospinal muscles

19. a. rectus capitus posterior major
    b. obliquus capitus superior
    c. obliquus capitus inferior

20. a. vertebral artery
    b. suboccipital nerve

21. C2

22. vertebral artery

23. L1 or L2

24. cauda equina

25. filum terminale (internum)

26. lumbar cistern

27. coccygeal ligament

28. denticulate ligament

29. a. superior margin of the iliac crest
    b. posterior margin of the external oblique
    c. anterior margin of the latissimus dorsi

30. nucleus pulposis

31. a. iliocostalis
    b. longissimus
    c. spinalis

32. a. semispinalis
    b. multifidus
    c. rotatores

33. a. cruciform ligament
    b. apical ligament
    c. alar ligament
    d. tectorial membrane

34. suboccipital nerve

35. conus medullaris

36. S2

2 Upper Limb

1. a. costocoracoid membrane
   b. suspensory ligament of the axilla

2. cephalic vein

3. cephalic vein

4. a. C8
   b. C6
   c. C7

5. median antebrachial veins

6. median cubital (antecubital) vein

7. musculocutaneous nerve

8. inferior border of the teres major

9. medial and lateral borders of the pectoralis minor

10. a. superior thoracic artery
    b. thoracoacromial artery and lateral thoracic artery
    c. subscapular artery
       anterior humeral circumflex artery
       posterior humeral circumflex artery

11. coracoid process

12. lateral lip of the intertubercular groove

13. anterior surface of the medial border of the scapula

14. lateral 1/3 of the clavicle, spine of the scapula, acromion process

15. floor of the intertubercular groove

16. lesser tuberosity of the humerus
17. deltoid tubercle of the humerus

18. subscapularis

19. serratus anterior

20. floor of the intertubercular groove

21. lateral border of the first rib

22. a. long head of the triceps
    b. lateral head of the triceps
    c. inferior margin of the teres minor
    d. superior margin of the teres major

23. a. axillary nerve
    b. posterior humeral circumflex artery

24. musculocutaneous nerve

25. radial nerve

26. radial nerve

27. a. radial artery
    b. ulnar artery

28. radial tuberosity

29. coronoid process of the ulna

30. median cubital (antecubital) vein

31. a. pronator teres
    b. brachioradialis

32. a. brachialis
    b. supinator

33. ulnar nerve

34. a. pronator teres
    b. flexor carpi radialis
    c. palmaris longus
    d. flexor carpi ulnaris

35. flexor digitorum superficialis

36. a. flexor digitorum profundus
    b. flexor pollicis longus
    c. pronator quadratus

37. flexor carpi ulnaris

38. palmaris longus

39. flexor digitorum profundus

40. a. supinator
    b. brachioradialis

41. radial nerve

42. a. abductor digiti minimi
    b. flexor digiti minimi
    c. opponens digiti minimi

43. a. abductor pollicis brevis
    b. flexor pollicis brevis
    c. opponens pollicis brevis
    d. adductor pollicis

44. flexor pollicis brevis

45. palmar interossei

46. dorsal interossei

47. ulnar nerve

48. superficial branch of the ulnar nerve

49. glenoid labrum

50. a. trapezoid ligament
    b. conoid ligament

51. coraco-acromial ligament

52. a. scaphoid
    b. lunate
    c. triquetrum
    d. pisiform

53. a. trapezium
    b. trapezoid
    c. capitate
    d. hamate

54. a. profunda brachii artery
    b. radial nerve

55. circumflex scapular artery

56. extensor pollicis longus

57. radial artery

58. scaphoid (navicular)

59. metacarpo-phalangeal joint (MC-P joint)

60. a. flexion
    b. extension
    c. extension

61. a. lateral head
    b. medial head

62. ulnar artery

63. posterior interosseous artery

64. a. superior ulnar collateral artery
    b. posterior ulnar recurrent artery

65. a. inferior ulnar collateral artery
    b. anterior ulnar recurrent artery

66. a. middle collateral artery
    b. interosseous recurrent artery

67. a. radial collateral artery
    b. radial recurrent artery

68. infero-medially from radius to ulna or supero-laterally from ulna to radius

69. scaphoid (navicular)

70. a. profunda (deep) brachial artery
    b. superficial ulnar collateral artery
    c. inferior ulnar collateral artery

71. a. ascending branch
    b. middle collateral artery
    c. radial collateral artery

72. suprascapular nerve

73. subscapularis

74. radius (radial tuberosity)

75. ulna (ulnar tuberosity)

76. a. scaphoid
    b. lunate

77. C5, C6, C7

78. a. upper subscapular nerve
    b. middle subscapular nerve (thoracodorsal nerve)
    c. lower subscapular nerve

79. adductor pollicis

80. flexor carpi ulnaris

81. pronator teres

82. supinator

3 Lower Limb

1. a. os coxae (innominate bone)
   b. sacrum

2. a. ilium
   b. ischium
   c. pubis

3. a. anterior superior iliac spine (ASIS)
   b. pubic tubercle

4. a. 126 degrees (range 115 – 140)

5. a. 7 degrees – males and 12 degrees – females

6. fascia lata

7. iliotibial band (tract)

8. cribriform fascia

9. femoral vein

10. popliteal vein

11. L2, L3, L4

12. L2, L3, L4

13. deep fibular (peroneal) nerve

14. anterior superior iliac spine

15. anterior inferior iliac spine

16. lesser trochanter

17. adductor magnus
18. femoral artery

19. a. inguinal ligament
    b. adductor longus
    c. sartorius

20. a. femoral vein
    b. femoral artery
    c. femoral nerve

21. medial

22. deep inguinal lymph nodes (node of Cloquet)

23. femoral nerve

24. saphenous nerve

25. lateral femoral cutaneous nerve

26. obturator nerve

27. profunda femoris artery

28. greater trochanter of the femur

29. a. sartorius
    b. gracilis
    c. semitendinosus

30. a. tibial nerve
    b. common fibular (peroneal) nerve

31. piriformis

32. tibial nerve

33. sural nerve

34. common fibular nerve

35. a. tibialis anterior
    b. extensor hallucis longus
    c. extensor digitorum
    d. fibularis tertius

36. anterior tibial nerve

37. a. gastrocnemius
    b. soleus
    c. plantaris

38. a. tibialis posterior
    b. flexor hallucis longus
    c. flexor digitorum longus
    d. popliteus

39. dorsalis pedis artery

40. a. plantar flexion
    b. eversion

41. superficial fibular nerve

42. a. gastrocnemius
    b. popliteus

43. calcaneus

44. saphenous nerve

45. tibial nerve

46. a. iliofemoral ligament
    b. ischiofemoral ligament
    c. pubofemoral ligament

47. obturator artery

48. profunda femoris aartery

49. posterior cruciate ligament

50. medial ligament of the ankle

51. posterior tibial artery

52.
   a. profunda (deep) femoral artery
   b. superficial circumflex iliac artery
   c. superficial external pudendal artery
   d. superficial epigastric artery
   e. deep external pudendal artery
   f. descending genicular artery

4 Thoracic Wall

1. a. manubrium
   b body
   c. xiphoid process

2. 2nd rib

3. ribs 1-7

4. T5

5. T4

6. a. innermost intercostal
   b. internal intercostal

7. a. internal thoracic artery
   b. aorta

8. right

9. a. vein
   b. artery
   c. nerve

10. a. depresses lower ribs
    b. elevates upper ribs

11. a. lateral thoracic artery
    b. internal thoracic artery

12. a. musculophrenic artery
    b. superior epigastric artery

13. T4/T5

14. 6th costal cartilage

5 Pleura and Lungs

1. a. costal pleura
   b. mediastinal pleura
   c. diaphragmatic pleura
   d. cervical pleura

2. azygos vein

3. arch of the aorta

4. pulmonary ligament

5. T4 – T5

6. a. accessory hemiazygos vein
   b. azygos v.

7. right

8. a. 8th rib
   b. 10th rib
   c. 12th rib

9. 6th costal cartilage

10. 4th costal cartilage

11. a. 6th rib
    b. 8th rib
    c. 10th rib

12. 6th costal cartilage

13. 4th costal cartilage

14. a. T3
    b. 5th
    c. 6th

15. a. 5th rib
    b. 4th

16. C3, C4, C5

17. right superior lobe

18. anterior scalene

19. anterior (the phrenic nerve lies anterior to the root of the lung)

20. pericardiophrenic artery and vein

6 Mediastinum

1. a. sternal angle (angle of Louis)
   b. T4-T5 intervertebral disc

2. a. anterior interventricular artery
   b. circumflex artery
   c. left marginal artery.
   d. lateral (diagonal)

3. a. posterior interventricular artery
   b. right marginal artery
   c. AV nodal branch
   d. SA nodal branch

4. tricuspid valve

5. a. intermediolateral nucleus.
   b. superior, middle, inferior cervical gang.
   c. dorsal motor nucleus
   d. cardiac plexus

6. a. brachiocephalic trunk
   b. left common carotid artery
   c. left subclavian artery

7. superior vena cava

8. right atrium

9. right pulmonary artery

10. anterior cardiac veins

11. esophagus

12. a. left internal jugular vein
    b. left subclavian vein

13. anterior

14. anterior
15. anterior

16. posterior

17. posterior

18. posterior

19. posterior

20. right

21. anterior

22. subclavian artery

23. vena caval hiatus

24. anterior

25. posterior

26. a. T12
    b. T8
    c. T10

27. the second intercostal space on the right side along the lateral sternal border

28. the fifth intercostal space on the left side along the lateral sternal border

29. the second intercostal space on the left side along the lateral sternal border

30. the fifth intercostal space on the left side on or near the midclavicular line

31. parietal layer of serous pericardium

32. right ventricle

33. right coronary artery

34. thoracic aorta

35. left pulmonary artery

7 Abdomen

1. transversalis fascia

2. transversalis fascia

3. a. inferior epigastric artery
   b. deep circumflex iliac artery

4. a. superficial circumflex iliac artery
   b. superficial epigastric artery

5. superior epigastric artery

6. T10

7. inferior epigastric artery

8. a. anterior superior iliac spine
   b. pubic tubercle

9. invagination of the transversalis fascia

10. opening in the aponeurosis of the external oblique

11. a. femoral vein
    b. femoral artery
    c. femoral nerve

12. a. external oblique
    b. internal oblique
    c. transversalis fascia

13. internal oblique

14. genital branch of the genitofemoral nerve

15. epididymis

16. peritoneum

17. a. inferior vena cava
    b. left renal vein

18. a. gastrophrenic ligament
    b. gastrosplenic ligament
    c. gastrocolic ligament

19. a. hepatogastric ligament
    b. hepatoduodenal ligament

20. a. hepatic artery
    b. portal vein
    c. bile duct

21. a. striated
    b. both striated and smooth
    c. smooth

22. a. vagus nerve
    b. vagus nerve
    c. vagus nerve

23. a. cardia
    b. fundus
    c. body
    d. pylorus

24. celiac artery

25.	a	right gastric artery	(common) hepatic artery
	b.	left gastric artery	celiac trunk
26.	a.	right gastroepiploic artery	gastroduodenal artery
	b.	left gastroepiploic artery	splenic artery

27. a. left gastric artery
    b. splenic artery
    c. (common) hepatic artery

28. a. right gastric artery
    b. gastroduodenal artery
    c. proper hepatic artery

29. a. right gastroepiploic artery
    b. superior pancreaticoduodenal artery

30. a. splenic vein
    b. superior mesenteric vein

31. a. cystic duct
    b. common hepatic duct

32. a. common bile duct
    b. main pancreatic duct (duct of Wirsung)

33. 2nd part (descending part) through the sphincter of Oddi

34. major duodenal papilla

35. accessory pancreatic duct (duct of Santorini)

36. L3

37. superior mesenteric artery

38. ligament of Treitz

39. parts 2 (descending), 3 (horizontal) and 4 (ascending)

40. 1st and 2nd parts

41. a. 3rd (horizontal) part of the duodenum
    b. transverse colon

42. descending colon and sigmoid colon

43. descending colon and sigmoid colon

44. ascending colon and descending colon

45. a. portal vein
    b. hepatic artery
    c. bile duct

46. falciform ligament

47. a. quadrate lobe
    b. caudate lobe

48. umbilical vein

49. fetal ductus venosus

50. posterior

51. anterior

52. a. inferior phrenic artery
    b. abdominal aorta
    c. renal artery

53. T5 – T8

54. T9 – T11

55. inferior epigastric vessels

56. obliterated umbilical artery

57. L4

58. L5

59. a. inferior pancreaticoduodenal artery
    b. intestinal arteries
    c. ileocolic artery
    d. right colic artery
    e. middle colic artery

60. a. left colic artery
    b. sigmoid arteries
    c. superior rectal artery

61. right and left lumbar lymphatic trunks

62.
- a. ilioinguinal nerve
- b. genital branch of the genitofemoral nerve

63. above
64. superior hypogastric plexus

65. ileum

66.
- a. teniae mesocolica
- b. teniae omentalis
- c. teniae libra

67. genitofemoral nerve

68.
- a superficial (fatty) layer (Camper's fascia)
- b. deep (membranous) layer (Scarpa's fascia)

69. external spermatic fascia

70. deep fascia of the penis (Buck's fascia)

71.
- a. iliohypogastric nerve
- b. ilioinguinal nerve
- c. genitofemoral nerve
- d. lateral femoral cutaneous nerve
- e. femoral nerve
- f. obturator nerve
- g. accessory obturator nerve
- h. lumbosacral trunk

72.
- a. lumbar veins
- b. right gonadal (testicular and ovarian) veins
- c. renal veins
- d. right suprarenal vein
- e. right inferior phrenic vein
- f. hepatic vein

73.
- a. splenic artery
- b. celiac artery

c. hepatic artery
d. splenic artery
e. gastroduodenal artery

74. a. psoas major
b. quadratus lumborum

75. a. spleen
b. stomach
c. pancreas

76. a. latissimus dorsi
b. external oblique
c. iliac crest

77. a. rectus abdominus
b. inferior epigastric vessels
c. inguinal ligament

78. ligamentum teres (round ligament)

8 Pelvis and Perineum

1. a. ilium
   b. ischium
   c. pubis

2. false pelvis

3. sacrotuberous ligament

4. sacrospinous ligament

5. a. levator ani
   b. coccygeus

6. a. puborectalis
   b. pubococcygeus
   c. iliococcygeus

7. piriformis

8. puborectalis

9. broad ligament (mesometrium)

10. levator ani

11. internal iliac a (ant. div.)

12. internal iliac (ant. div.)

13. common iliac vein

14. pudendal nerve

15. S2 – S4

16. S2 – S4

17. hypogastric nerves

18. anterior

19. detrusor muscle

20. smooth muscle

21. striated (skeletal) muscle

22. pelvic splanchnic nerves

23. a. ureteric orifices
    b. urethral orifice

24. a. inferior mesenteric artery
    b. internal iliac artery
    c. internal pudendal artery

25. a. ductus deferens
    b. seminal glands

26. bulbourethral (Cowper's) glands

27. a. infundibulum
    b. ampulla
    c. isthmus
    d. uterine part

28. broad ligament

29. a. body
    b. cervix

30. a. inferior vena cava
    b. left renal vein

31. a. inferior vena cava
    b. left renal vein

32. a. perimetrium

b. myometrium
c. endometrium

33. vesicouterine pouch

34. rectouterine pouch

35. a. urogenital triangle
b. anal triangle

36. a. inferior fascia of the pelvic diaphragm
b. perineal membrane

37. deep transverse perineus

38. a. bulbospongiosus
b. ischiocavernosus
c. superficial transverse perineus

39. obturator internus

40. smooth muscle

41. striated muscle

42. pectinate line

43. pectinate line

44. corpus spongiosum

45. corpus cavernosum

46. a. ilioinguinal nerve
b. genital branch of the genitofemoral nerve
c. superficial branch of the pudendal nerve
d. perineal branch of the posterior femoral cutaneous nerve

47. internal pudendal artery

48. vestibular (Bartholin's) glands

49. broad ligament (mesosalpinx)

50. a. inferior epigastric artery
    b. deep circumflex iliac artery

51. a. iliolumbar artery
    b. lateral sacral artery
    c. superior gluteal artery
    d. obturator artery
    e. internal pudendal artery
    f. inferior gluteal artery
    g. umbilical artery
    h. inferior vesical artery
    i. middle rectal artery
    j. (female) uterine artery
    k. (female) vaginal artery

52. a. inferior mesenteric vein
    b. internal iliac vein
    c. internal iliac vein

53. a. prostatic urethra
    b. membranous urethra
    c. spongy urethra
    d. penile urethra

54. tunica albiguinea

55. pampiniform plexus

56. ductus deferens

57. a. sphincter urethrae
    b. deep transverse perineus

58. pudendal nerve

59. greater vestibular glands

60. inferior rectal nerve

9 Head and Face

1. a. occipital bone
   b. parietal bone

2. a. frontal bone
   b. parietal bone

3. a. zygomatic bone
   b. temporal bone

4. a. frontal bone
   b. parietal bone
   c. temporal bone
   d. sphenoid bone

5. approximately 18 months

6. falx cerebri

7. tentorium cerebelli

8. a. skin
   b. dense connective tissue
   c. epicranial aponeurosis
   d. loose connective tissue
   e. periosteum (pericranuim)

9. a. facial artery
   b. lingual artery
   c. superficial temporal artery
   d. superior thyroid artery
   e. maxillary artery

10. facial artery

11. a. superior sagittal sinus
    b. inferior sagittal sinus

c. superior petrosal sinus
d. cavernous sinus
e. straight sinus

12. a. inferior petrosal sinus
b. sigmoid sinus

13. vertebral artery

14. middle meningeal artery

15. a. right anterior cerebral artery
b. left anterior cerebral artery

16. a. middle cerebral artery
b. posterior cerebral artery

17. a. ophthalmic artery
b. anterior choroidal artery
c. posterior communicating artery
d. middle cerebral artery
e. anterior cerebral artery

18. facial nerve

19. a. temporalis
b. masseter
c. lateral pterygoid
d. medial pterygoid

20. a. ophthalmic nerve
b. maxillary nerve
c. mandibular nerve

21. glossopharyngeal nerve
a. otic ganglion

22. facial nerve
a. submandibular ganglion

23. facial nerve
a. pterygopalatine ganglion

24. facial nerve

25. trigeminal nerve

26. trigeminal nerve
    a. ophthalmic nerve
        1) superior orbital fissure
    b. maxillary nerve
        1) foramen rotundum
    c. mandibular nerve
        1) foramen ovale

27. jugular foramen

28. stylopharyngeus

29. a. taste from posterior 1/3 of tongue
    b. arterial pressure measured at the carotid sinus

30. jugular foramen

31. a. laryngeal muscles
    b. upper third of the pharyngeal constrictors

32. slowing of heart rate

33. increased motility

34. a. blood gas monitoring at carotid body
    b. taste from the epiglottis

35. a. foramen magnum
    b. jugular foramen

36. a sternocleidomastoid
    b. trapezius

37. hypoglossal canal

38. genioglossus

39. facial nerve

40. internal auditory (acoustic) meatus

41. stylomastoid foramen

42. a. temporal branch
    b. zygomatic branch
    c. buccal branch
    d. marginal mandibular branch
    e. cervical branch

43. a. greater superficial petrosal nerve
    b. deep petrosal nerve

44. facial nerve (nervus intermedius)

45. facial nerve (nervus intermedius)

46. glossopharyngeal nerve

47. a. trigeminal
    b. vagus

48. a. tensor tympani
    b. tensor veli palatini
    c. anterior belly of the digastric
    d. mylohyoid

49. a. posterior belly of the digastric
    b. stapedius
    c. stylohyoid

50. superior orbital fissure

51. foramen rotundum

52. foramen ovale

53. optic foramen

54. a. facial nerve
    b. vestibular nerve
    c. cochlear nerve

55. inferior alveolar nerve

56. ciliary ganglion

57. pterygopalatine ganglion

58. otic ganglion

59. submandibular ganglion

60. maxillary nerve

61. internal carotid artery

62. abducens nerve

63. nasopalatine nerve

64. maxillary artery

65. facial nerve

66. glossopharyngeal nerve

67. glossopharyngeal nerve

68. otic ganglion

69. mental nerve

70. frontal bone

71. ethmoid bone

72. sphenoid bone

73. temporal bone

74. a. sphenoid bone
    b. occipital bone

75. emissary veins

76. diploic veins

77. a. great vein of Galen
    b. inferior sagittal sinus

78. chorda tympani

10 Mouth and Pharynx

1. a. masseter
   b. temporalis

2. mylohyoid

3. palatoglossus
   a. vagus nerve

4. palatopharyngeus
   a. vagus nerve

5. palatoglossus

6. palatopharyngeus

7. glossopharyngeal nerve

8. a. levator veli palatine
   b. tensor veli palatine

9. vagus nerve

10. hypoglossal nerve

11. hypoglossal nerve

12. palatoglossus
    a. vagus nerve

13. trigeminal nerve

14. glossopharyngeal nerve

15. glossopharyngeal nerve

16. glossopharyngeal nerve

17. facial nerve
18. glossopharyngeal nerve

19. maxillary nerve (trigeminal V2)

20. mandibular nerve (trigeminal V3)

21. greater palatine nerve

22. lesser palatine nerve

23. foramen cecum

24.
    a. vagus nerve
    b. vagus nerve
    c. glossopharyngeal nerve

25. palatine tonsil

26. facial nerve

27.
    a. pterygomandibular raphe
    b. hyoid bone
    c. thyroid cartilage

28.
    a. stylopharyngeus
    b. palatopharyngeus
    c. salpingopharyngeus

29. eustation (auditory or pharyngotympanic) tube

30. retropharyngeal space

11 Neck and Larynx

1. a. investing fascia
   b. pretracheal fascia
   c. prevertebral fascia
   d. carotid sheath

2. a. sternocleidomastoid
   b. trapezius

3. a. infrahyoid muscles
   b. thyroid gland
   c. trachea
   d. esophagus

4. a. carotid arteries
   b. internal jugular vein
   c. vagus nerve

5. a. midline of neck
   b. anterior border of sternocleidomastoid
   c. inferior border of mandible

6. a. anterior border of sternocleidomastoid
   b. posterior belly of digastric
   c. superior belly of omohyoid

7. a. posterior belly of digastric
   b. anterior belly of digastric
   c. inferior border of mandible

8. a. anterior belly of digastric
   b. hyoid bone
   c. midline beneath mandible

9. a. superior belly of omohyoid
   b. anterior border of sternocleidomastoid
   c. sternohyoid

10. a. anterior border of trapezius
    b. posterior border of sternocleidomastoid
    c. superior border of clavicle

11. a. sternohyoid
    b. thyrohyoid
    c. sternothyroid
    d. omohyoid

12. a. digastric
    b. stylohyoid
    c. mylohyoid
    d. geniohyoid

13. posterior cricoarytenoid

14. internal laryngeal nerve (branch of superior laryngeal)

15. a. lateral cricoarytenoid
    b. transverse arytenoid

16. internal laryngeal nerve

17. cricothyroid

18. external laryngeal nerve

19. internal laryngeal nerve

20. a. anterior scalene
    b. middle scalene

21. posterior

22. anterior

23. lateral border of first rib

24. a. vertebral artery
    b. thyrocervical trunk (artery)
    c. internal thoracic artery

d. costocervical trunk (artery)
e. dorsal scapular artery

25. external carotid artery
26. common carotid artery

27. a. superficial temporal vein
b. maxillary vein

28. a. retromandibular vein
b. posterior auricular vein

29. a. internal jugular vein
b. subclavian vein

30. C3, C4, C5

31. C1, C2, C3

32. C4

33. C1 - C4

34. C5 and C6

35. C7, C8, (T1)

36. a. ascending pharyngeal artery
b. occipital artery
c. posterior auricular artery
d. superior thyroid artery
e. lingual artery
f. facial artery

37. a. maxillary artery
b. superficial temporal artery

38. a. inferior thyroid artery
b. transverse cervical artery
c. suprascapular artery

39. a. lesser occipital nerve
    b. great auricular nerve
    c. transverse cervical nerve
    d. supraclavicular nerve(s)

40. anterior scalene

41. a. internal jugular vein
    b. internal jugular vein
    c. left brachiocephalic vein

42. a. occipital triangle
    b. omoclavicular triangle

43. a. anterior border of trapezius
    b. posterior border of sternocleidomastoid
    c. inferior belly of omohyoid

44. a. posterior border of sternocleidomastoid
    b. inferior belly of omohyoid
    c. superior border of the clavicle

45. sternocleidomastoid

46. anterior scalene

47. spinal accessory nerve (XI)

48. greater occipital nerve

49. a. chemoreceptors
    b. baroreceptors

50. a. brachiocephalic artery (trunk)
    b. left vertebral artery
    c. left subclavian artery

51. phrenic nerve

52. a. right subclavian artery
    b. arch of the aorta or ligamentum arteriosum

53. a. thyrocervical trunk
    b. external carotid artery

54. a. internal jugular vein
    b. internal jugular vein
    c. brachiocephalic vein

55. C4

56. cricothyroid membrane

57. thyrohyoid membrane

58. cricothyroid muscle

59. a. internal laryngeal nerve (branch of the superior laryngeal nerve)
    b. superior thyroid artery

60. a. internal laryngeal nerve
    b. inferior laryngeal nerve

61. thyrohyoid membrane

62. arytenoid cartilage

63. posterior atlantooccipital membrane

64. facial nerve

65. 2nd, 3rd and 4th

66. a. sigmoid sinus
    b. inferior petrosal sinus
    c. lingual vein
    d. pharyngeal veins
    e. facial vein
    f. superior thyroid vein
    g. middle thyroid vein

67. a. meningeal branches
    b. auricular branch

68. inferior pharyngeal constrictor

69. internal laryngeal nerve

70. facial artery

71. omoclavicular triangle (of the posterior triangle)

72. occipital triangle (of the posterior triangle)

73. roots of the brachial plexus

74. middle and inferior cervical ganglia (stellate)

75. a. thyroid cartilage
    b. cricoid cartilage
    c. arytenoid cartilage
    d. corniculate cartilage
    e. cuneiform cartilage
    f. epiglottic cartilage

76. a. vestibular folds
    b. vocal fold

77. vocalis

78. C7

79. great auricular nerve

80. a. lesser occipital nerve
    b. spinal accessory nerve
    c. supraclavicular nerve
    d. transverse cervical nerve
    e. great auricular nerve

81. dorsal scapular nerve

82. a. transverse cervical artery
    b. suprascapular artery

83. anterior

84. sternothyroid muscle

12 Orbit

1. a. optic canal
   b. superior orbital fissure
   c. inferior orbital fissure

2. a. optic nerve
   b. ophthalmic artery

3. a. frontal nerve
   b. lacrimal nerve
   c. nasociliary nerve
   d. oculomotor nerve
   e. trochlear nerve
   f. abducens nerve
   g. superior ophthalmic vein
   h. inferior ophthalmic vein

4. a. oculomotor nerve (superior branch)
   b. oculomotor nerve (inferior branch)
   c. nasociliary nerve
   d. abducens nerve

5. conjunctiva

6. a. levator palpebrae superioris
   b. superior tarsal muscle (Mueller's muscle)

7. a. tarsal glands
   b. ciliary glands

8. anterior in the supero-lateral margin

9. below the inferior nasal concha

10. facial nerve

11. inferior oblique

12. oculomotor

13. superior cervical ganglion

14. a. frontal nerve
    b. nasociliary nerve
    c. lacrimal nerve

15. maxillary nerve

16. nasociliary nerve

17. nasociliary nerve

18. ophthalmic artery

19. cavernous sinus

13 Eye

1. a. sclera (fibrous)
   b. choroid (vascular)
   c. retina (neural)

2. long ciliary > nasociliary > trigeminal

3. iris

4. aqueous humor

5. aqueous humor

6. a. pupillary constrictor — parasympathetic
   b. pupillary dilator — sympathetic

7. suspensory ligament of the lens (zonular fibers of Zinn)

8. accommodation of the lens

9. Edinger-Westphal nucleus — ciliary ganglion

10. Edinger-Westphal nucleus — ciliary ganglion

11. intermediolateral nucleus — superior cervical ganglion

12. ciliary body

13. canal of Schlemm (scleral venous sinus)

14. lateral margin of the anterior chamber

15. point of exit of retinal ganglion cell axons from the retina

16. center of the optic disc

17. retinal region of high visual acuity

18. center of macula lutea – area of highest visual acuity

19. optic disc

20. a. rods
    b. cones

21. cone

22. ophthalmic artery

23. a. cavernous sinus (posteriorly)
    b. pterygoid sinus (inferiorly)
    c. angular vein (anteriorly)

14 Ear

1. a. great auricular nerve
   b. auriculotemporal nerve

2. malleus 1st arch
   incus 2nd arch
   stapes 3rd arch

3. malleus

4. stapes

5. tensor tympani
   a. trigeminal

6. stapedius
   a. facial

7. cochlear (auditory)

8. internal auditory meatus

9. semicircular canals

10. utricle and saccule (otolithic organs)

11. perilymph

12. perilymph

13. endolymph

14. endolymph

15. utricle and saccule (otolithic organs)

16. spiral ganglion

17. vestibular (Scarpa's) ganglion

18. anterior-inferior quadrant

19. a. chorda tympani
    b. lesser superficial petrosal nerve

20. 2nd branchial arch

21. 1st branchial arch

About the Authors

Michael F. Nolan is professor of Basic Science Education at the Virginia Tech Carilion School of Medicine in Roanoke. He received his Physical Therapy training at Marquette University and his PhD in Human Anatomy from the Medical College of Wisconsin. Nolan spent the first 34 years of his career teaching gross anatomy and neuroanatomy to medical students and resident physicians at the University of South Florida. He has received more than 20 awards for excellence in teaching including the Master Teacher Award in 2014 from the International Association of Medical Science Educators and the John M. Thompson Outstanding Teacher Award in Neurosurgery in 2006. He has published over 40 peer-reviewed articles and book chapters as well as four textbooks in human gross anatomy and neuroanatomy.

John P. McNamara is the Director of Anatomy and Assistant Professor of Basic Science Education at the Virginia Tech Carilion School of Medicine in Roanoke. His doctoral training is in chiropractic from Life University (Marietta, GA) with undergraduate (Lock Haven University of Pennsylvania) and graduate (Shippensburg University of Pennsylvania) degrees. He is also ABD from Virginia Tech in Educational Leadership and Policy Studies. For nearly the past 30 years, McNamara has maintained a private practice in Salem, VA, and taught full-time anatomy and physiology, gross anatomy, neuroanatomy, and pathophysiology at the College of Health Sciences (Jefferson College) in Roanoke. From 2013 to 2017 he taught the gross anatomy course for the Doctor of Physical Therapy program at Radford University in Roanoke. He is licensed to practice as a Doctor of Chiropractic in both Virginia and Pennsylvania, and he is certified as an Emergency Medical Technician in Virginia.

Made in United States
Orlando, FL
25 August 2022

21529564R00096